1542

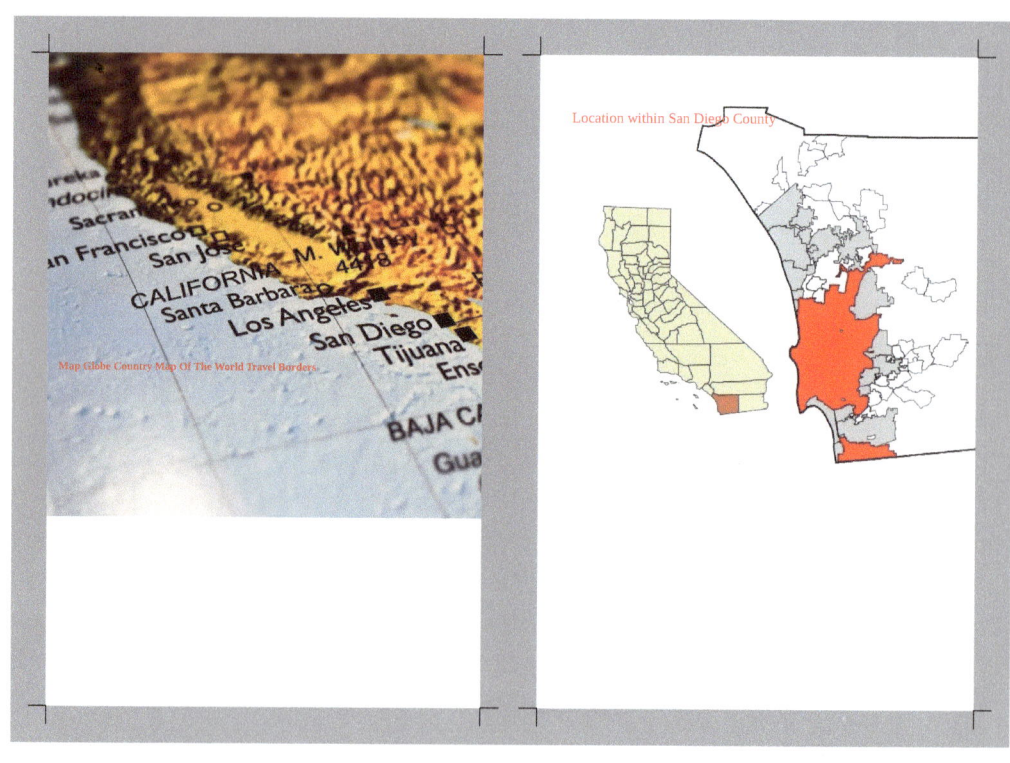

Sand Beach Barefoot Standing Gap Wake Water

Urban aerial of San Diego and Tijuana, Mexico

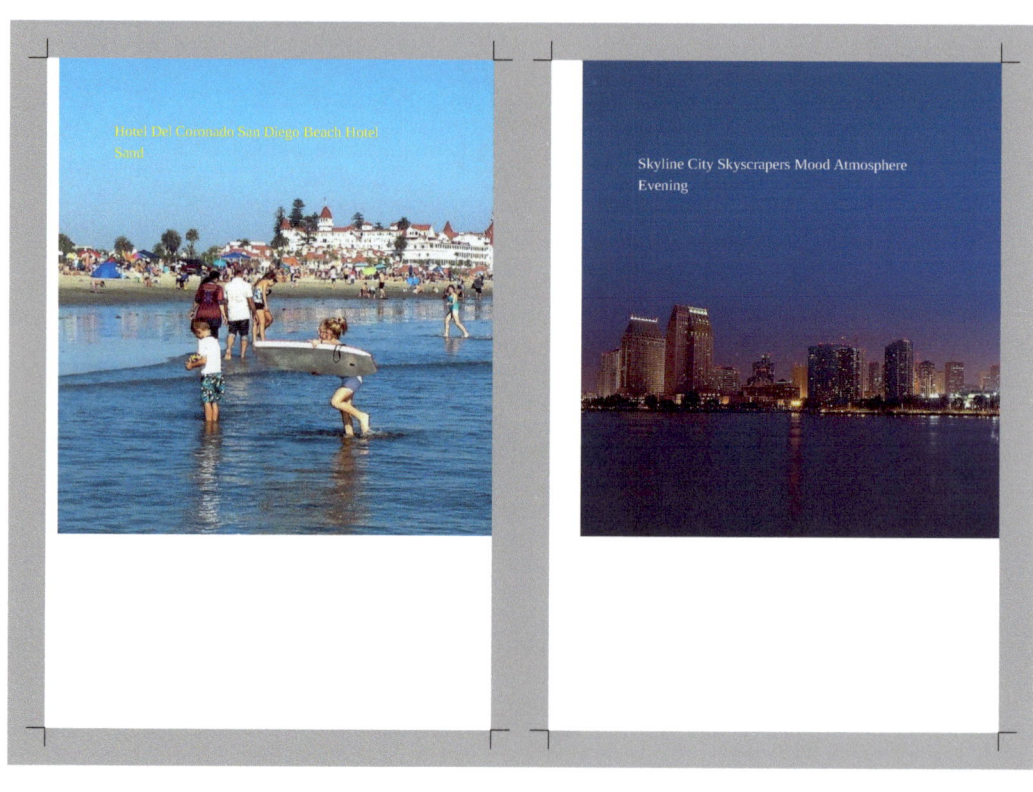

Evening Night Lights Long Exposure Sea Port City

Petco Ball Park san Diego California Aerial View

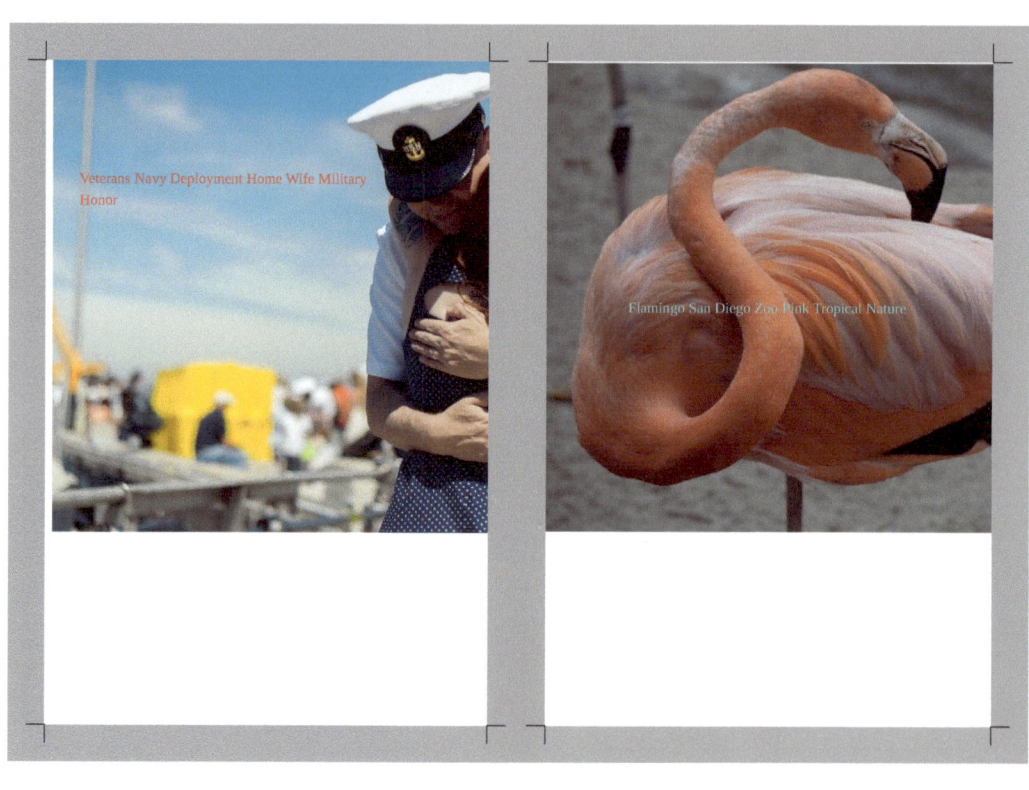

San Diego Harbor Dr Convention Center

Baby Gorilla Baby San Diego Zoo

Sea Lions San Diego La Jolla Animal Wildlife

Veterans Navy Deployment Home Mother Mom Son

Palm Trees Cactus Old Town Market San Diego USA

San Diego Mission Mission San Diego California

San Diego California Cycling Bicyclist Bicycle

San Diego California Ocean Sky Buildings Water

San Diego California Uss Carl Vinson Navy Sky

Elephants Baby Elephant Zoo San Diego Zoo Animal

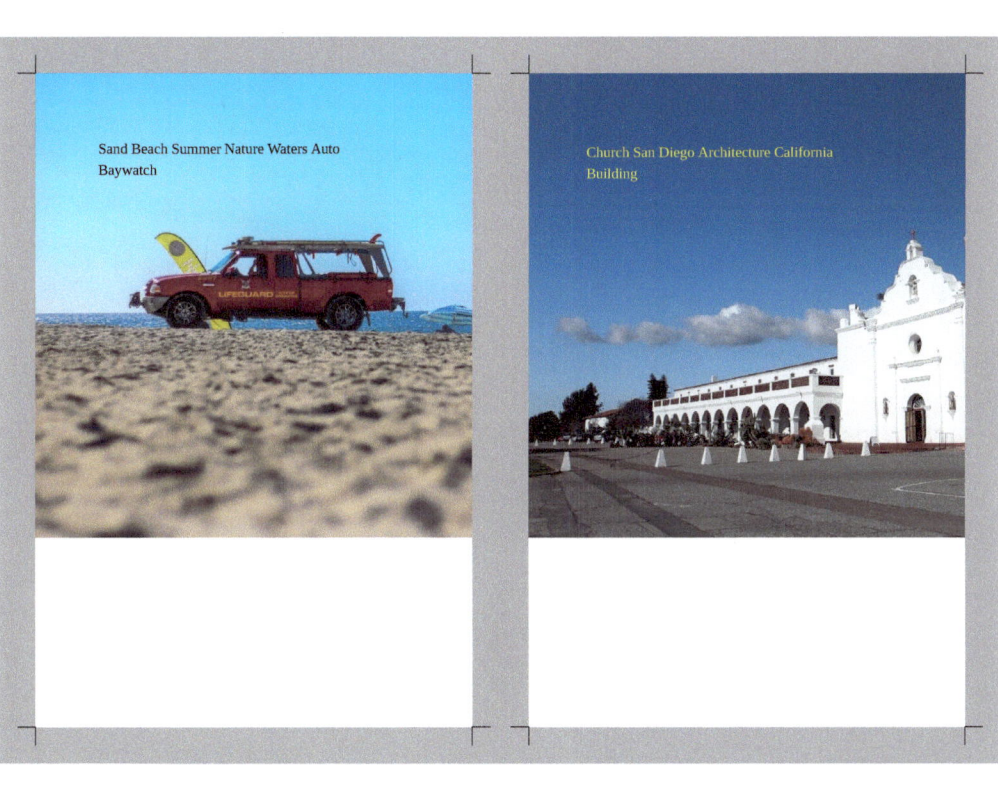

Bird Safari Park San Diego

California Dock Pier Coast Pacific Usa San Diego

Beach Estate San Diego Architecture California

Port Hamburg Museum Ship Cap San Diego

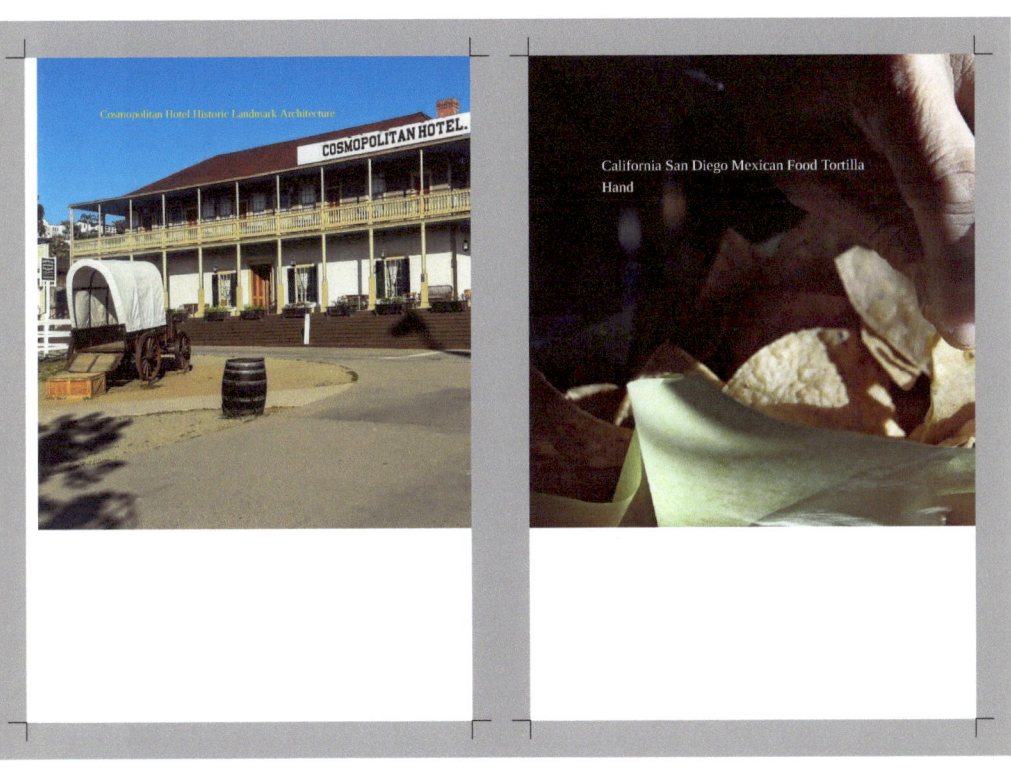

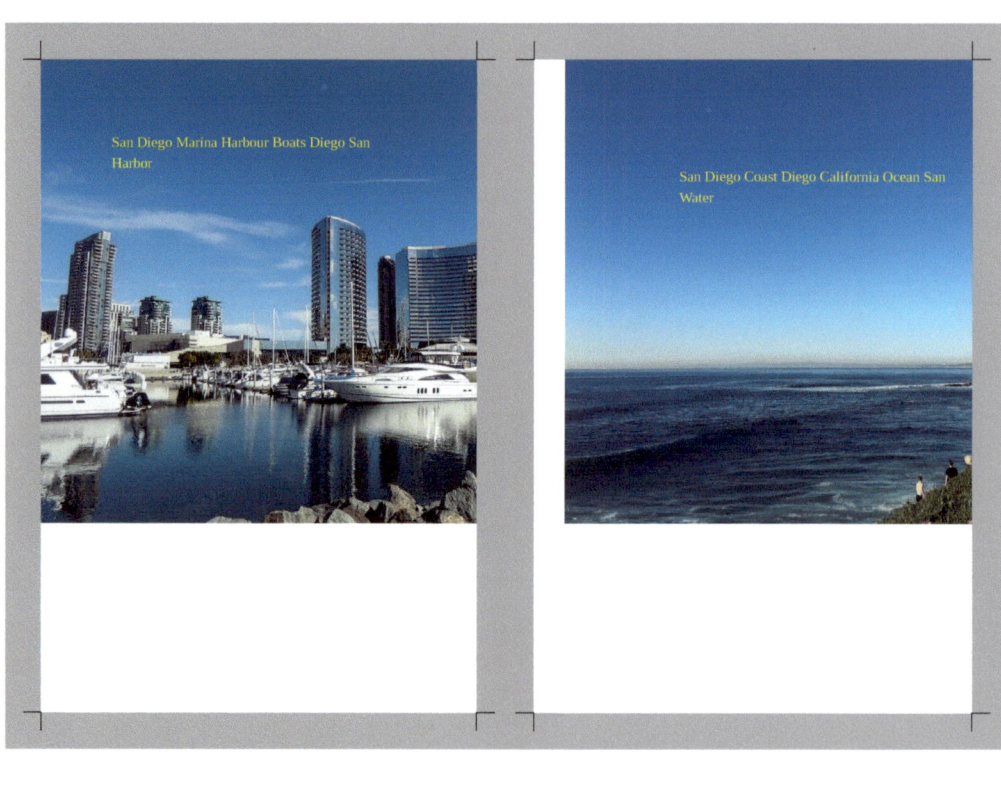

San Diego De Alcala Mission California Adobe White

Sea World Dolphins Water Marine Jump Aquarium

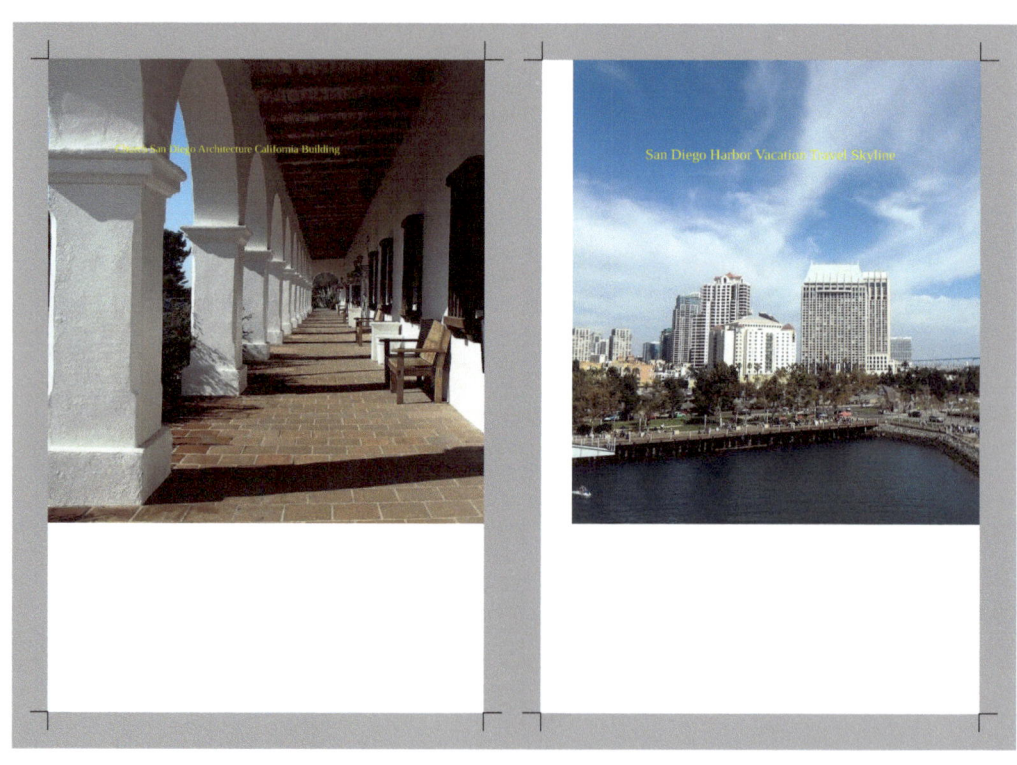

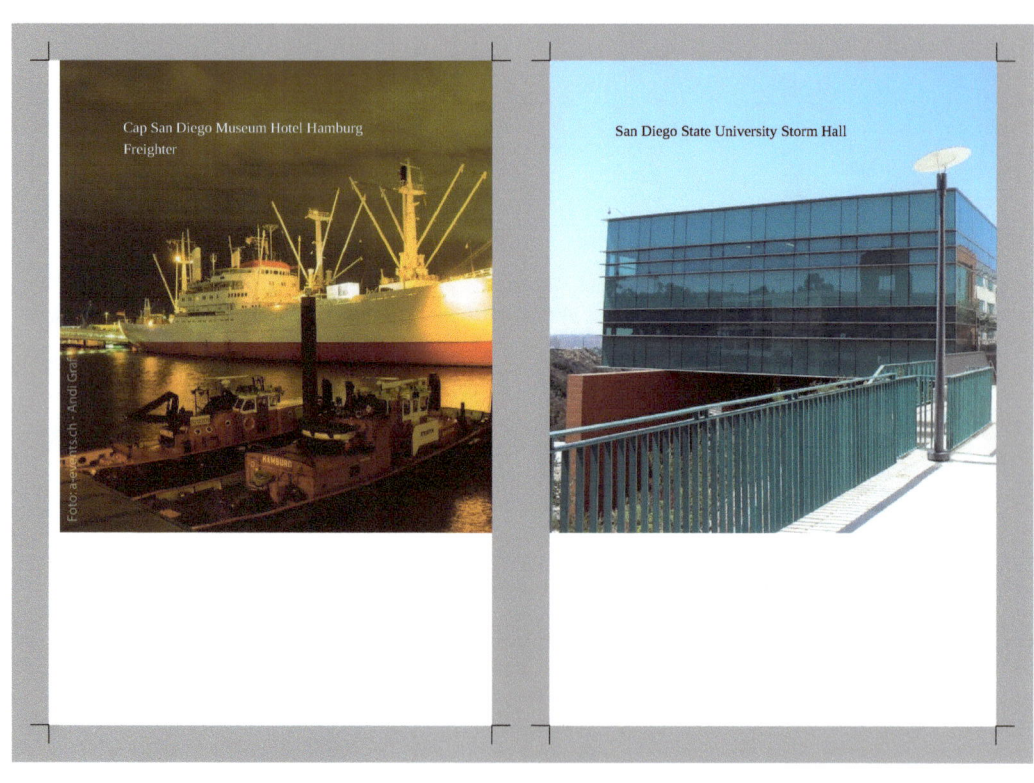

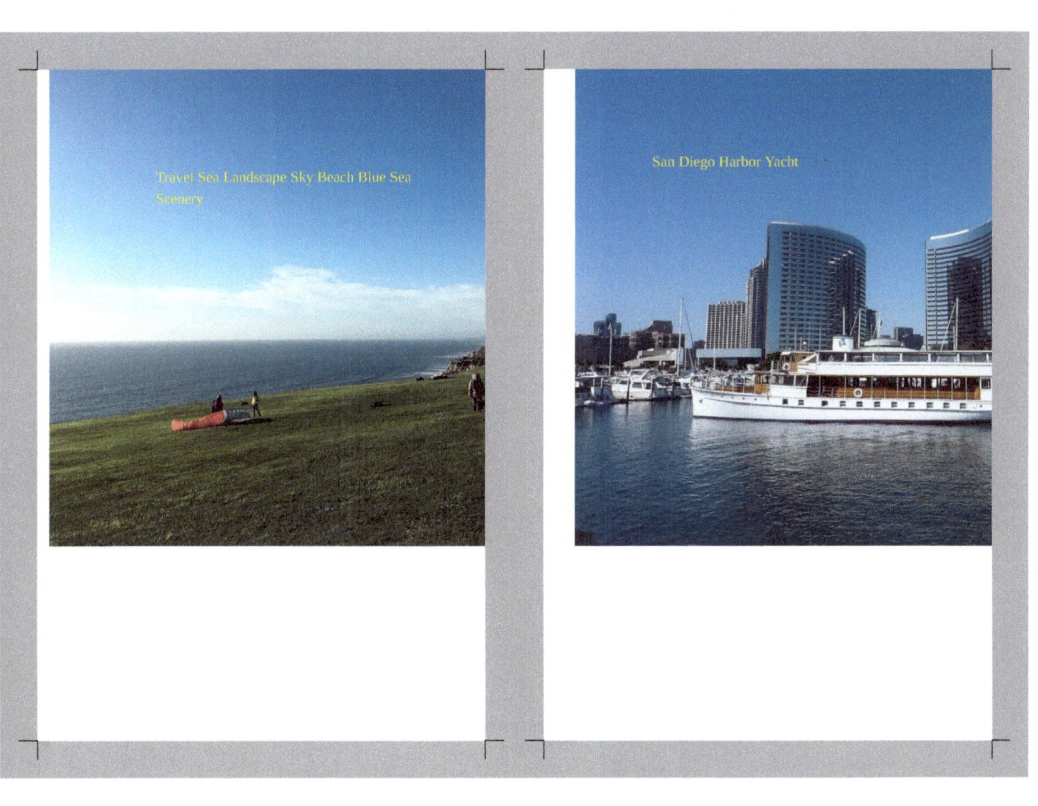

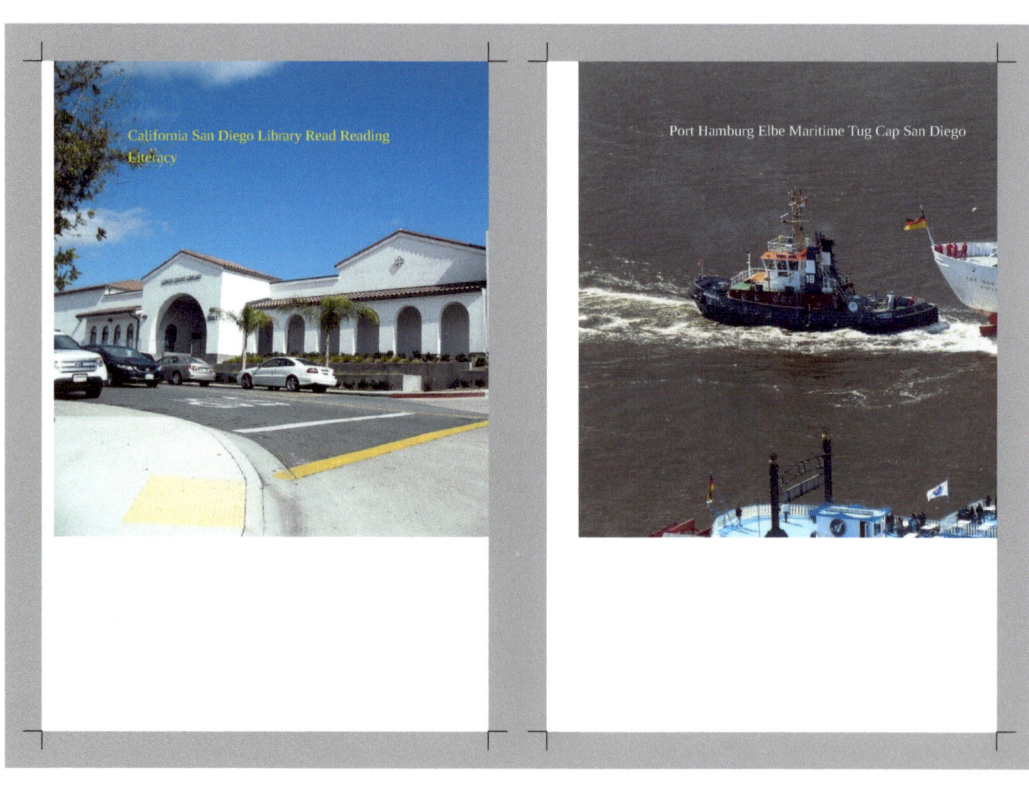

Proof

www.ingramcontent.com/pod-product-compliance
Lightning Source LLC
Chambersburg PA
CBHW040316220526
45473CB00009B/2456